Oh no, not Moe Mosquito!

By Delta Hatcher

Copyright © 2015 by Delta Hatcher, a pen name for the BIOL 413 (Parasitology) class of Christian Brothers University, which included the following individuals:

Cale Alexander
Keyara Baltimore
Jacob Beaty
John Buttross
Chelsea Casaccia
Justin Crone (guest illustrator)
Rachel Depperschmidt
Daria Dyer
Kyle Fioranelli
Daniel Gabriel
Patrick Gurley
Leslie Hogan
Joseph Krebs
Jacob Mann
Brent McGlaughlin
Juan Mejia
Katie Robinson
Angelica Rodriguez (not pictured at right)
Daniel Schenck
Bailey Smith
Camisha Terrell
Ryan Tomlinson

Oh no, not Moe Mosquito!

Table of Contents:
I. Biology of mosquitoes
II. The diseases transmitted by mosquitoes
III. Control and prevention

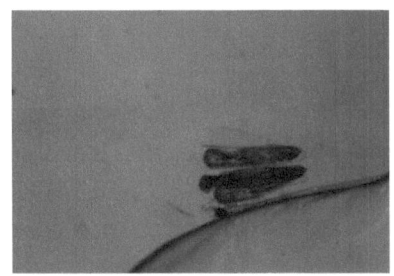
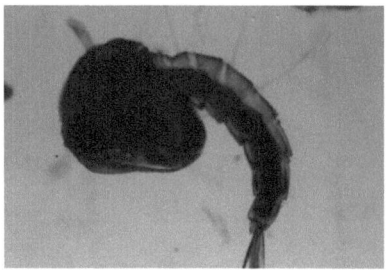
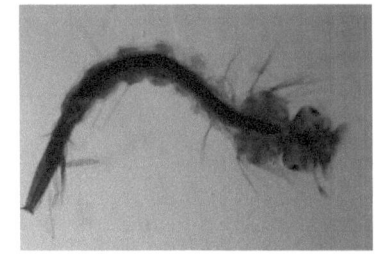
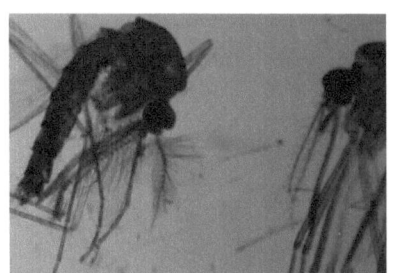

Oh, no, not Moe Mosquito! Copyright © 2015 by Delta Hatcher.

All rights reserved. Printed in the United States of America by CreateSpace, an imprint of Amazon. No part of this book may be used or reproduced in any manner whatsoever without written permission. For information, contact Dr. Stan Eisen, Professor and Director, Preprofessional Health Programs, Christian Brothers University, 650 East Parkway South, Memphis, TN 38104.

ISBN-13:
978-1519518910

ISBN-10:
1519518919

I. Biology of Mosquitoes

Hello! My name is Monique, but my friends call me Moe. I am a mosquito, and I am here to teach you all about mosquitoes, the diseases they cause, and how you can keep them from making you sick.

Depending on a mosquito's species, she lays eggs in one of two places. She will lay them either on damp soil or on open water, like in a ditch. Those eggs that are laid in water will immediately hatch, but those laid on damp soil will remain dormant until the soil gets thoroughly wet after a good rainstorm, and that gives the egg the signal to hatch, and enough time for the larvae to grow.

If she is a mosquito who needs a permanent water source she will lay her eggs on standing water. And the eggs must stay moist. They cannot withstand drying. If the eggs dry out, they die and larvae will never hatch. If everything goes as planned the eggs will hatch within twenty-four hours.

How she lays her eggs is a big deal too! She can lay up to 300 eggs at once! If she lays a lot of eggs at once she usually lays them in clusters called **egg rafts**.

    Once the egg has hatched it has passed the baby stage. He's not enclosed in the dark alone within that tight space anymore. He is now a larva! Being in this stage of life, he can finally stretch. The new larva spends most of his time on the surface of water like some of the other larva and occasionally travels.

    The newly formed larva stays on the water's surface to breathe unlike his friend, Phil. Phil must stay parallel to the water surface to intake oxygen. Being the larva of a mosquito means he is able to live in water but still must take in oxygen from air. For gas exchange, he has a siphon tube which allows him to hang upside down at the water's surface and accumulates oxygen from his tail. He swims around to find food like algae or microorganisms. The cool part about being a larva is the advancing levels. Each mosquito goes through four stages called **instars** which are mini-puberty changes where they elongate. Completing these stages means the larva can advance to the **pupa** stage.

The third stage of mosquito development is known as the "pupa stage". This stage is similar to a toddler (the larva) growing older and becoming a kid (the pupa form). It lives in water, where it inhales air through a snorkel-like device called an "air siphon". The pupa air siphon is similar to the larval form, but it is on the pupa's back instead of at the end of his tail. The pupa is one of the laziest life stages.

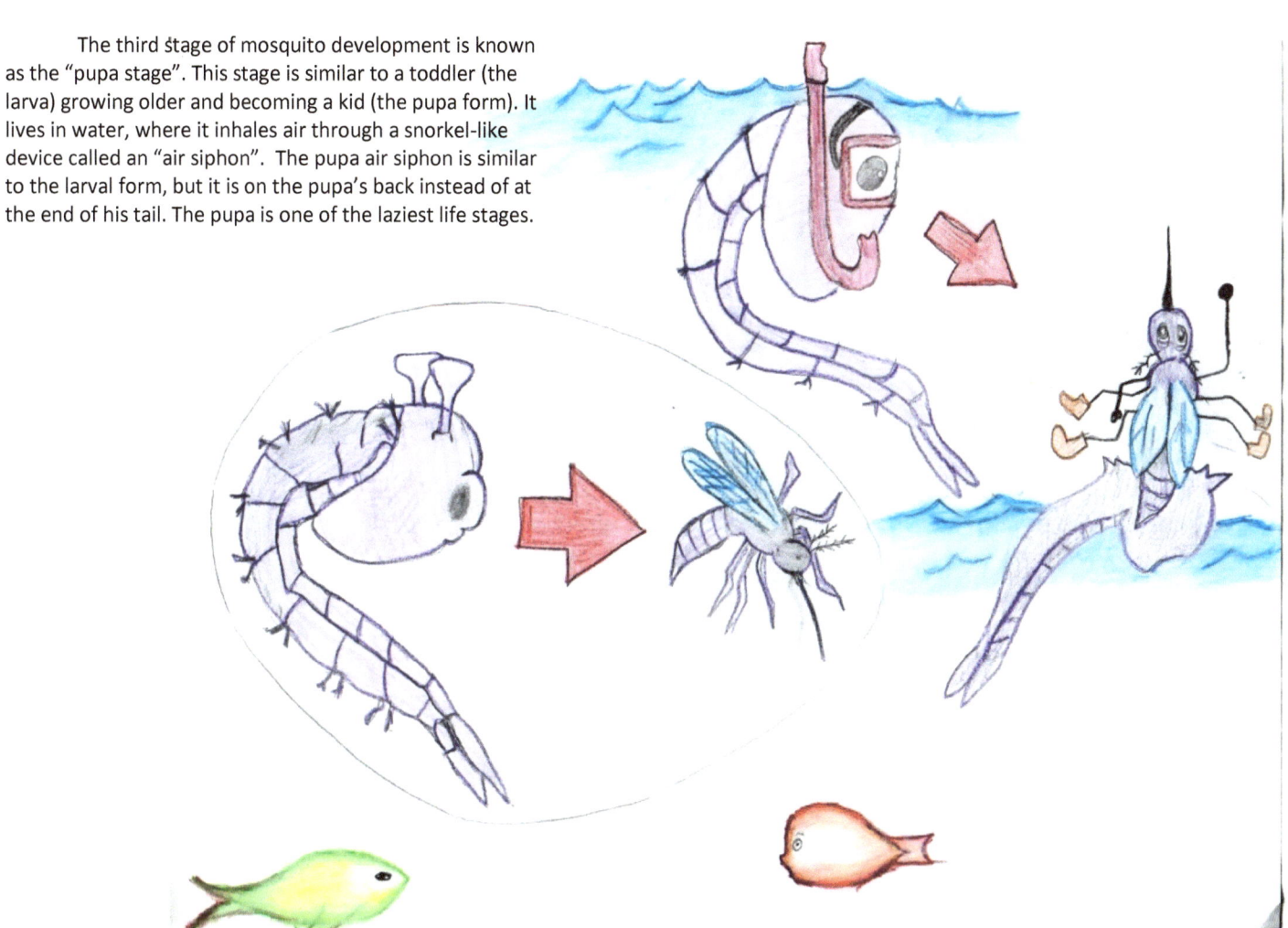

In fact, it is such a lazy stage that it does not even eat any food! Instead the pupa is encased inside a shell and undergoes internal changes to turn into an adult. After a few to several days, an adult mosquito emerges from of the pupa mosquito, as if the pupa is a piñata and the adult is a giant candy bar inside.

After hatching from the pupae, the adults male and female are free to fly away.

The adult males and females mate to make new eggs. The process, however, is not so simple.

Adult mosquitoes, especially in Memphis, sometimes only live for 5-10 days, so there isn't much time for dating! The male mosquito, especially adapted to quickly find the female, has large feathery antennas that are used to detect her fragrance like a fine perfume. Once mating is completed, the female must take a blood meal to produce eggs. That's right, only the female bites! After she has taken human blood she is able to lay more eggs, starting the entire process again.

However, what may be a quick meal for the female could leave lasting problems for the human...

II. Diseases transmitted by Mosquitoes

We girl mosquitoes *love* blood. In fact, we *need* it to live and reproduce. We like to snack on blood from all sorts of animals including you juicy humans. One day, we might eat some nice bird blood, and the next day, we might decide to go out for people or dog the next time we are hungry. Since a lot of diseases are carried by organisms in blood, we often pick up a parasite, bacterium, or virus in one of our blood meals, and then share it with whoever is on the menu next.

Here's some more detail on the diseases we can transmit:

West Nile Virus

West Nile virus originated in the West Nile sub-region of eastern Uganda, but today is common worldwide, including in all of the continental United States and Canada. Some of my cousins really have a taste for birds, and that is how West Nile is spread around. One of them bites a virus-infected bird. My cousins then carry the virus around for a couple of days and then when they bite one of you or another animal they pass it along in their saliva.

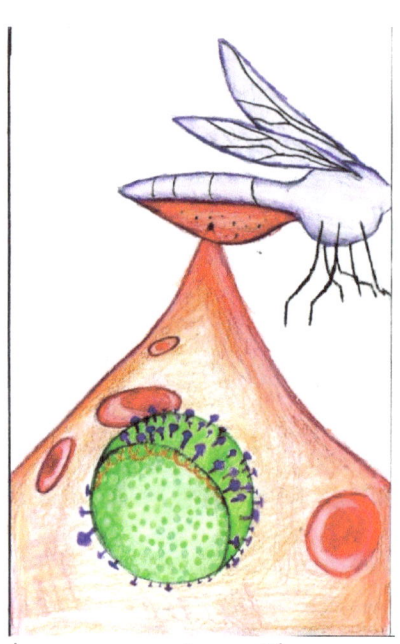

Some people who get West Nile Virus or WNV, about 80% of those infected, don't show any symptoms. The primary symptoms in the other 20% are headache, body aches, and vomiting. If the disease is not treated, a more severe form of illness may develop and progress to meningitis or West Nile Encephalitis. The more severe symptoms include high fever, disorientation, convulsions, and paralysis. It is approximated that only one in 150 people infected with WNV will develop the more severe form. Treatment is supportive, and may include hospitalization, intravenous fluid, respiratory support, and treatment for secondary infections.

Eastern Equine Encephalitis (EEE)

Eastern Equine Encephalitis (EEE) is a viral disease that is mainly carried by birds. And of course, mosquitos like birds too, which is where we pick up the virus. We often pass the virus along to horses, which is where the "equine" part of the name comes from, and we can give it to humans too, but not very often. EEE can make both horses and humans sick, but it isn't contagious among them. That means you can't give EEE to other kids by coughing or sneezing on them.

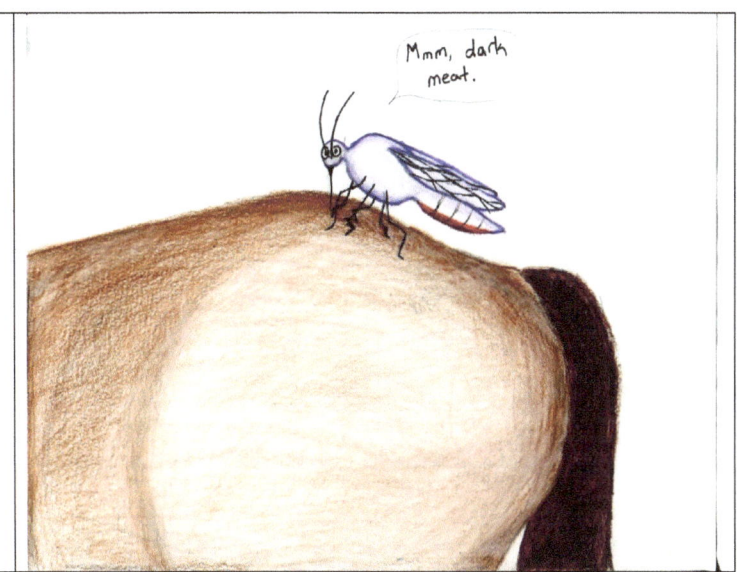

Symptoms are similar to those of WNV and may last one to two weeks. Most infected people have mild symptoms or none at all. As with WNV, a few patients may develop encephalitis, which can make them severely ill. Symptoms include fever, seizures, and coma. Treatment includes fluid, rest, and medicine to treat the symptoms. Encephalitis will require hospitalization.

And, yes, that *is* a picture of a horse's behind…

LaCrosse Encephalitis (LACV)

LaCrosse Encephalitis (LACV) is a virus spread by my pal *Aedes triseriatus*. He likes to hang out in woods or forests and finds chipmunks and squirrels especially tasty, but will also dine on human blood when people wander into the woods. Like EEE and WNV, most people get lucky and don't feel sick. People with weakened immune systems and children under sixteen are more likely to become ill. Fever, headache, and nausea are the initial symptoms. In a few infected patients, the virus invades the neurological system and can cause encephalitis, seizures, and coma. That will require treatment in a hospital. The milder cases just need rest, fluids, and medication to ease symptoms.

Chikungunya Disease

Encephalitis isn't the only kind of disease we like to share. I'd like to introduce you to my new friend Chikungunya. He's a virus originally from Africa and Asia and is new to the United States. He was being carried by people traveling from his home, but is now being carried around Florida by my cousins there.

Even though he's a little guy, he can do a lot of damage. If he decides he wants to be part of your life, you may not know it for about a week. Once you start to show symptoms, he'll hang around for another week or so. While you are sick, he'll give you a rash, make your head hurt, and even make your muscles and joints sore. For a very few people, he makes their joints ache long after he's gone.

As long as you drink plenty of fluids, get a lot of rest, and take medicine to relieve your headache and joint pain, you will be back on your feet in no time.

Heartworm (Dirofilaria immitis)

If you have a dog or a cat, mosquitoes can also make them sick by transmitting heartworms, roundworms which live *in* the bloodstream, on the right side of the heart. Mosquitoes pick up microscopic larvae from a dog that is already infected. We then transmit the worms by biting a pet dog or cat. The heartworm larvae enter the pet's skin through the bite wound. Still very tiny, the worms wiggle their way through the pet's bloodstream, and travel to the right side of the heart. On the way, the worms grow, sometimes up to 12 inches in length.

Heartworms are large enough to interfere with heart and lung function, and can cause your pet to cough or have trouble breathing. You can imagine that the heart cannot work very well when it is clogged up with worms! Your pet may become very weak and tired.

Once the worms become adults, it is difficult to get rid of them. Your dog may have to stay at the vet's office for several weeks while undergoing treatment. If the heartworm infection is severe, your pet may have permanent damage to the heart and lungs.

The good news is that heartworm infection is easily treated when the worms are still very tiny larvae, so you can prevent your pet from becoming sick from them. Your veterinarian will prescribe medicine to be given on the same day every month. Most varieties come in a form that looks and tastes like a treat so your pet will want to eat it.

III. Control

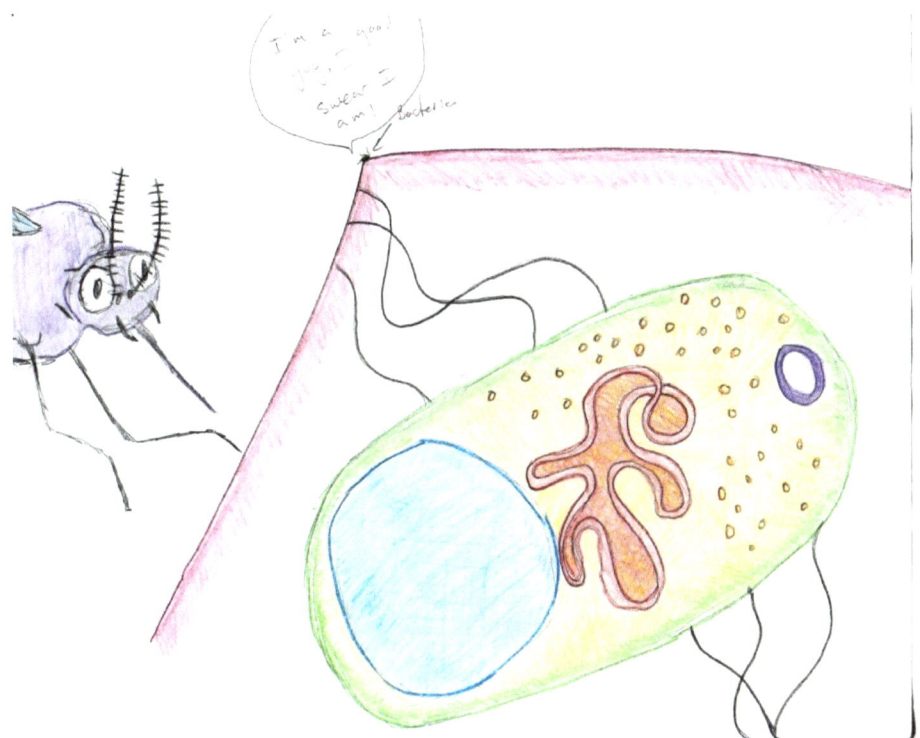

One way to keep mosquitos from making people sick is to keep them from becoming adults. This can be done by killing the mosquitos before they grow up. *Bacillus thuringiensis* is a tiny bacteria that kills mosquito larva.

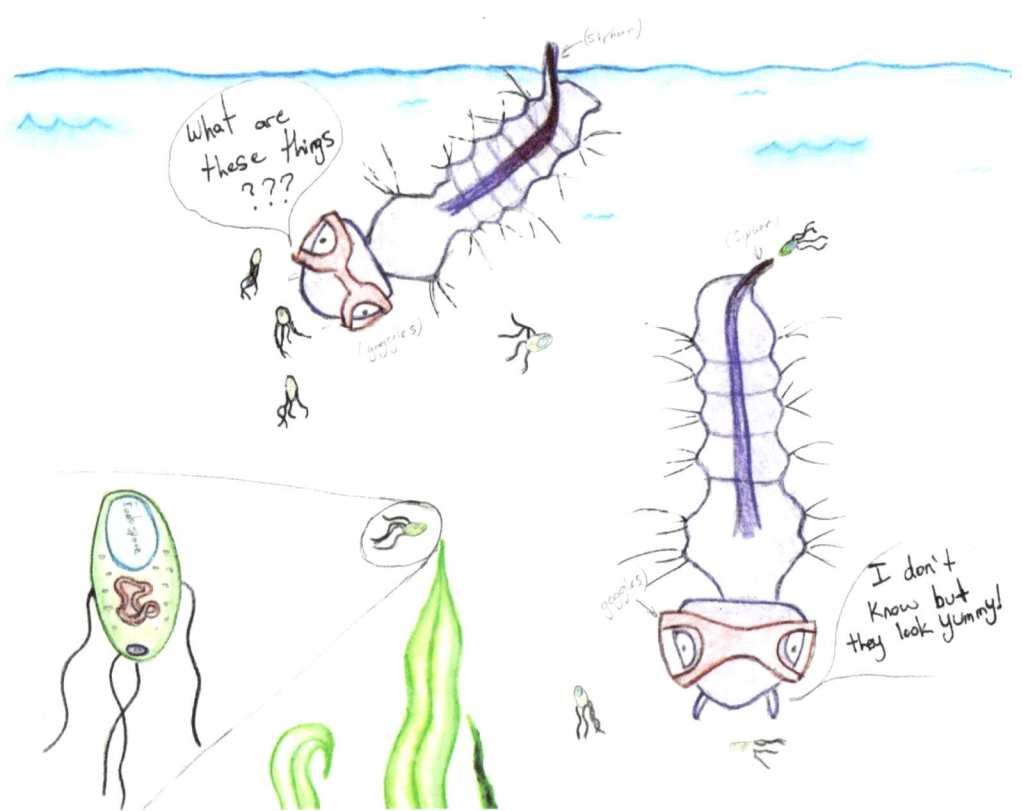

This bacteria comes in a little pellet that you can put into the water where the mosquito larvae are living. The bacteria release a chemical that acts as a poison to the larvae, causing them to bleed on the inside. Within a few days, most of the mosquito larvae will be dead.

Another method commonly used to control mosquito larvae is the use of light oil that can be applied directly to water via a backpack sprayer. Oils work by forming films on the surface of bodies of water. Under normal circumstances, the surface tension of water is strong enough to support the weight of the mosquito larvae; however, this isn't the case when oil has been applied. Oil alters the surface tension of water such that mosquito larvae struggle to make any contact with air above the water's surface.

Eventually, the oil descends into the air siphon and causes blockage. When this happens, the larvae lack the ability to derive oxygen from the air and as a result suffocate.

Juvenile Hormone prevents the larval mosquito from developing any further, so it stays in the larval stage. As a result, it never grows into a blood-sucking adult.

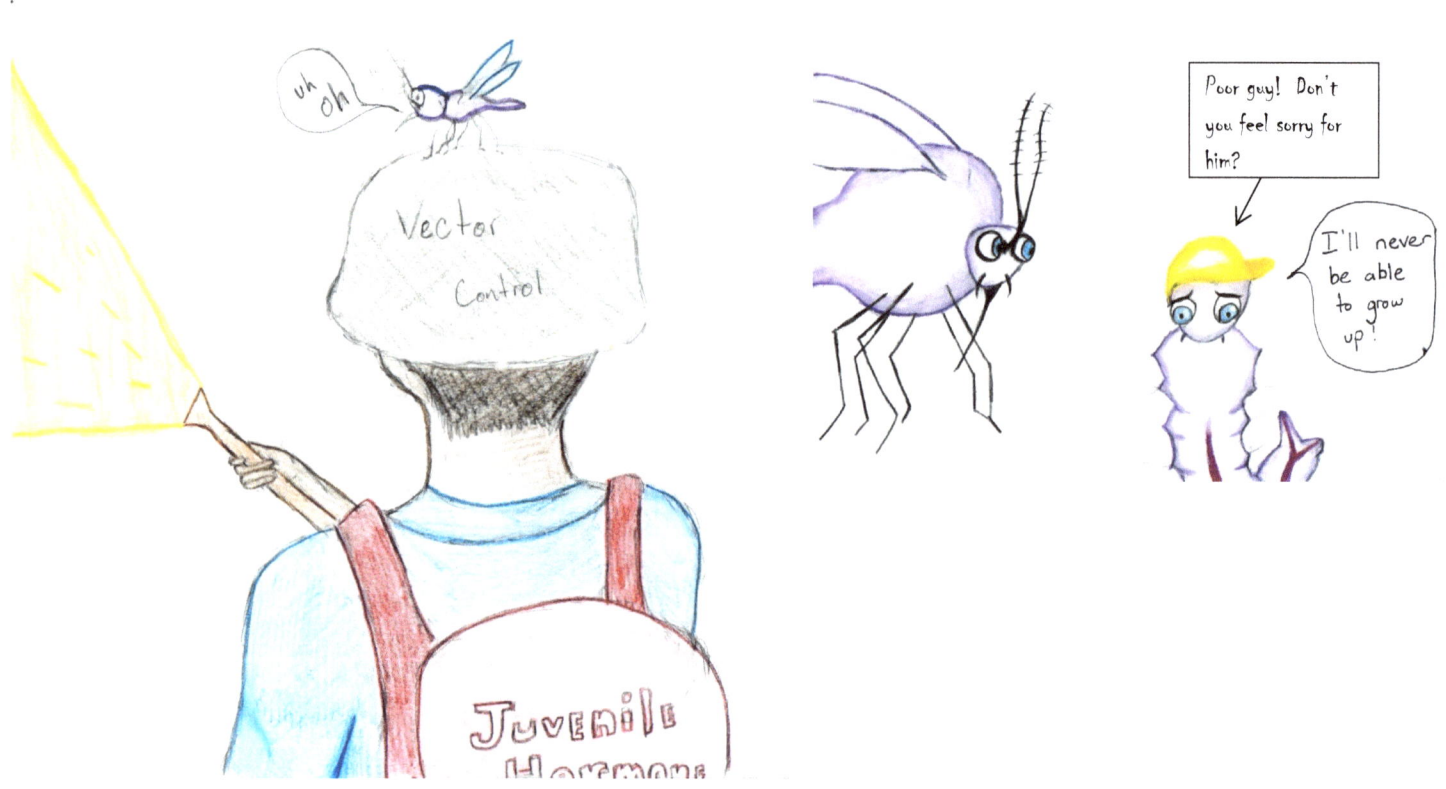

When city workers try to control the number of adult mosquitos like myself, they use chemical sprays called insecticides. Insecticides are chemicals used to kill adult mosquitos, but they are harmless to people and animals. These insecticides are usually dispersed throughout neighborhoods by vector control trucks.

Another effective way to control adult mosquito populations is to drain swamps and clear out marshy areas that commonly provide a beneficial environment for mosquitos to live and reproduce. This technique is useful because it prevents adult female mosquitos from laying eggs in these particular areas.

As we've already learned, mosquitos can be quite harmful creatures despite how small we may be. Even with all of the hard work that public health care officials do to control the mosquito population, there are still many of us out there looking to bite you.

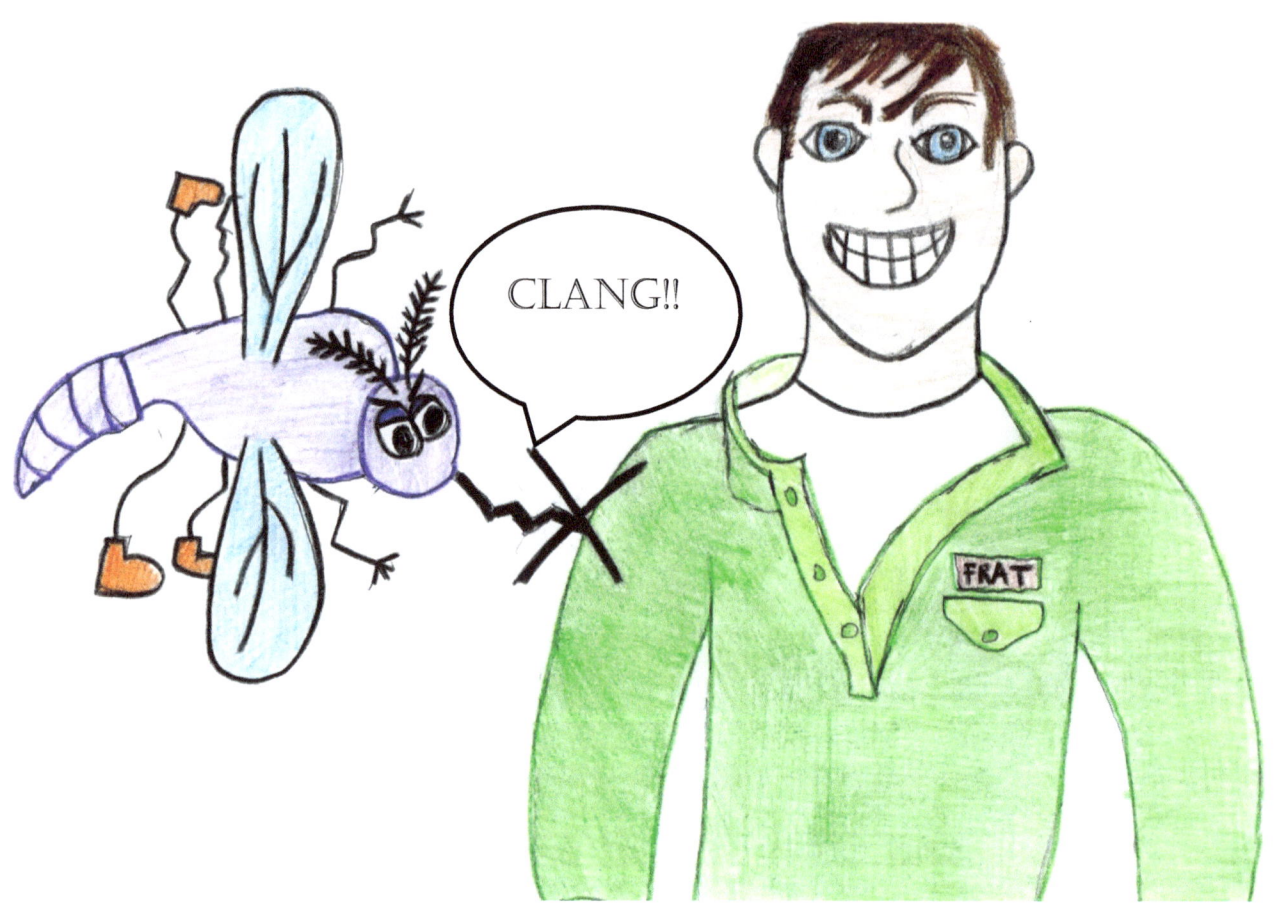

Luckily, there are some things you can do to prevent mosquitos from biting you. When working outdoors you should always remember to wear long sleeved shirts and pants to minimize the chances of being bitten.

Furthermore, mosquito repellants should be worn so that adult mosquitos will stay far away from you and your friends. Mosquito repellants are chemicals that are unpleasant to mosquitos but pose no threat to humans. This leads to no bites from mosquitos, no disease transmission, and no itchy red bumps! Remember to reapply your mosquito repellant every few hours while you're outside, otherwise the bug spray wears off and you may be bitten!

www.ingramcontent.com/pod-product-compliance
Lightning Source LLC
Chambersburg PA
CBHW041320180526
45172CB00004B/1168